Continents Activity Book

Author . . . . Kathy Rogers
Cover Design . . . . Denice McAskin

Table of Contents

Pages		Pages	
2	Continents	23-24	Asia
3-4	North America	25	Find Out About Asia
5	Know North America	26	Mount Fuji
6-7	Ancient Cultures	27-28	Australia
8-9	South America	29	Around Australia
10-11	Rain Forest Life	30	The Great Barrier Reef
12	Exploring South America	31-32	Antarctica
13-14	Europe	33	Amazing Antarctica
15	European Landmarks	34	Antarctic Food Web
16-17	Countries of Europe	35-37	Cuisine from the Continents
18-19	Africa	38-39	Games Around the World
20-21	Countries of Africa	40-48	Animals of the World
22	The Sahara Desert		

METRIC CONVERSION CHART

Refer to this chart when metric conversions are not found within the activity.

¼ tsp.	= 1.25 ml	¼ cup	= 60 ml	350° F	= 175° C	1 inch	= 2.54 cm
½ tsp.	= 2.5 ml	⅓ cup	= 75 ml	375° F	= 190° C	1 foot	= 30 cm
1 tsp.	= 5 ml	½ cup	= 125 ml	400° F	= 200° C	1 yard	= 91 cm
1 Tbsp.	= 15 ml	1 cup	= 230 ml	425° F	= 220° C	1 mile	= 1.6 km
		1 oz.	= 28 g				
		1 lb.	= .45 kg				

Reproducible for classroom use only.
Not for use by an entire school or school system.
EP093 • ©2004 Edupress, Inc.™ • P.O. Box 883 • Dana Point, CA 92629
www.edupressinc.com
ISBN 1-56472-093-4
Printed in USA

Continents

Astronauts traveling in space look back at planet Earth and see it in a way that most people never can. What they see is a huge sphere, about two-thirds of which is covered with water. They also see large areas of dry land, which are called continents. There are seven continents on the planet Earth: Asia, Africa, Europe, North America, South America, Antarctica, and Australia. Australia is sometimes grouped with New Zealand and other islands in the Pacific, and called Oceania. There are four large oceans that surround the continents: the Atlantic, Pacific, Indian, and Arctic Oceans. The seas surrounding Antarctica are sometimes considered a fifth ocean, the Southern Ocean.

Each of the seven continents has its own characteristics: landforms, plant and animal life, and cultures. Five of the continents are divided into smaller units that we know as individual countries. Australia is a continent and a country. The continent of Antarctica is uninhabited, except by scientists who visit from other parts of the world.

Project

Create a bulletin board to represent the continents and oceans of the world.

You may prefer to purchase our ready-to-use **Continents Bulletin Board, EP2231.**

Directions

1. Reproduce the map pages. Color as desired and cut out.
2. Cover bulletin board with blue paper to represent the oceans of the world.
3. Using the atlas or world map as a guide, place the continent cutouts in their correct positions.
4. Use index cards to create labels for the continents and major oceans of the world.
5. Use the bulletin board as a reference for your continent study.

Materials

- Atlas or world map
- Continent map patterns, pages 4, 9, 14, 19, 24, 28, and 32
- Blue bulletin board paper
- Index cards
- Markers or crayons
- Black marker for labels
- Scissors
- Push pins or tape

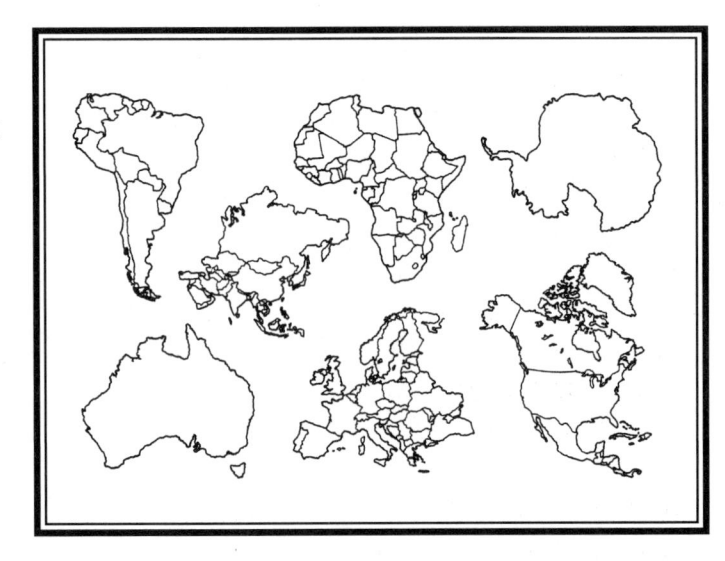

North America

The continent of North America lies between the Pacific and Atlantic Oceans, stretching from the Arctic Circle to tropical lands near the equator. North America is made up of Canada, the United States of America, Mexico, and the countries of Central America. The continent also includes Greenland, the French islands of St.-Pierre and Miquelon, and the Caribbean Islands.

North America is the third-largest continent in the world. Because of its size, the climate and geography from region to region can be very different. The northern lands of Canada and the upper part of the United States are arctic tundra. Further south, the land is covered with forests, then woodlands and prairies in the central regions. There are hot deserts in the southwest part of the continent, and a massive mountain range—the Rocky Mountains—runs from north to south near the western edge. The southernmost part of the continent is covered with tropical rain forests.

Project

Create a map showing some of the main geographic features of North America.

Materials

- Atlas, encyclopedia, or topographical map of North America
- Map page, following
- Pen or pencil
- Colored pencils

Directions

Using the atlas, encyclopedia, or map as a guide, find the location of the following geographic features and mark them on the map page. Be sure to label the features as you add them.

- Identify the Atlantic Ocean, the Pacific Ocean, and the Arctic Ocean.
- Locate the Gulf of Mexico and the Caribbean Sea.
- Find the three largest countries on the North American continent: Canada, the United States, and Mexico. Color Canada red. Color the United States yellow. Color Mexico green.
- Find and label the following Central American countries: Guatemala, Belize, Honduras, El Salvador, Nicaragua, Costa Rica, and Panama. Color each of them a different color.
- Find and label the five Great Lakes that lie on the border between Canada and the United States. Color them blue.
- Using a blue crayon, trace the course of the Mississippi River in the United States.
- Look for the island of Greenland and outline it in blue.

Continents Activity Book — ©Edupress, Inc.™ EP093

Map of North America

Know North America

North America is the third-largest continent in the world. Complete the crossword puzzle below to learn some facts about this continent. You will find the words you need to complete the puzzle in the box to the right. Use resource books or the Internet to help you find the answers.

Arctic	Hudson
Canada	Lake Superior
Central	McKinley
fifty	Mississippi
Greenland	Rio Grande
Great Lakes	thirteen

Across

2. The group of countries between Mexico and South America is known as _____ America.
3. The tallest mountain in North America is Mt. _____.
4. The northern part of Canada lies within the _____ circle.
6. Canada's _____ Bay is the largest bay in the world.
7. The five lakes that lie on the border between Canada and the United States are called the _____. (two words)
9. The longest river in North America is the _____ River.
11. This is the world's biggest island.

Down

1. The largest lake in North America is _____. (two words)
2. The three largest countries on the North American continent are Mexico, the United States, and _____.
5. The total number of provinces and territories in Canada
8. The river that forms part of the border between Mexico and the United States is known in the United States as the _____. (two words)
10. The number of individual states in the United States

Ancient Cultures

When explorers from Europe arrived in Mexico, they were astounded to find evidence of two advanced civilizations having lived in the land they meant to conquer: the Maya and the Aztecs. Although neither race would survive the arrival of Europeans, the influences of their cultures can be found in Mexico today.

Where great Mayan and Aztec cities and places of worship once existed, there are now great stone ruins. One of the most important Aztec objects that has survived to this day is a calendar stone called the Stone of the Sun. The circular stone is four feet (1.22 m) thick and 13 feet (4 m) in diameter. It weighs more than 24 tons (21.7 metric tons)! The outer edge of the stone is carved with the symbols for the 20 days of the Aztec calendar. In the center is the face of the sun god, with a protruding knife blade for a tongue. Other carvings on the stone are religious symbols related to the worship of this powerful god.

This stone was not just a calendar. It told when the world was supposed to have begun and when it would end. The Aztecs believed they were living in the fifth and final era. They believed that the four previous worlds had been destroyed by jaguars, hurricanes, volcanic fires, and torrential rains.

Project

- Study the detailed carvings of the Stone of the Sun while coloring the designs.
- Get a feel for its size with some measuring activities.

Materials

- Pattern page, following
- Colored pencils
- Tape measure

Directions

1. Reproduce the pattern to be used as a coloring page.
2. Try to identify the symbols for the Aztec months.
3. Look near the center of the stone. Try to identify the jaguars, hurricanes, fires, and rains—the destroyers of the earlier worlds.

How Big is Big?

1. Measure your height and the heights of some friends. How many of you could fit head-to-toe across the Stone of the Sun? Lie on the ground to demonstrate the size of this massive sculpture.
2. How much is 24 tons? If the average family car weighs about 4,000 pounds (1,814 kg), figure out how many cars you would need to equal the weight of the Stone of the Sun.

Stone of the Sun

Aztec Symbols for the Months

Rain	Flower	Crocodile	Wind
House	Lizard	Serpent	Death's Head
Deer	Rabbit	Water	Dog
Monkey	Grass	Reed	Jaguar
Eagle	Vulture	Motion	Flint Knife

South America

South America is the world's fourth-largest continent. It is linked to North America by a narrow strip of land called the Isthmus of Panama. South America stretches from the tropics surrounding the Caribbean Sea in the northeast to the cold waters of Cape Horn in the south.

The Andes mountains run down the west side of the continent. The mountains include plateaus, glaciers, volcanos, and some of the highest peaks in the Western Hemisphere. To the west of the mountains is a strip of humid coastal lowlands and dry desert. To the east lies the basin of the Amazon River, home of the world's biggest rain forest. The southern end of the continent narrows and is covered with fertile farmlands and rolling grasslands. The very tip of the continent, *Tierra del Fuego,* is a maze of rocky islands and channels.

While the northern and central parts of the continent are warm and rainy, the southern end is barren and cold.

Project

Create a map showing some of the main geographic features of South America.

Materials

- Atlas, encyclopedia, or topographical map of South America
- Map page, following
- Pen or pencil
- Colored pencils

Directions

Using the atlas, encyclopedia, or map as a guide, find the location of the following geographic features and mark them on the map page. Be sure to label the features as you add them.

- Identify the Atlantic Ocean, the Pacific Ocean, and the Caribbean Sea.
- Find the spot where the Isthmus of Panama connects South America to North America. Put an X on the spot.
- Brazil is the largest country in South America. Color it green.
- The Amazon River, the longest river in the world, is in Brazil. Trace the river in blue.
- Three small countries lie north of Brazil on the Caribbean coast: French Guiana, Suriname, and Guyana. Label these countries and color them yellow.
- The Andes mountains begin in Venezuela and run down the western edge of South America, ending at the southern tip of the continent. Draw mountain peaks to show the location of the Andes.

- Label each of the countries in the northern Andes region: Venezuela, Colombia, Peru, Ecuador, and Bolivia. Color each of them a different color.
- Chile is a long narrow country that lies on the western coast of South America. Color it red.
- Uruguay and Paraguay are countries where the land is suitable for raising cattle. Draw a cow in each of these countries.
- Argentina is a large agricultural nation, with farming in the northeast and central regions, and cattle and sheep ranching in the south. Color Argentina brown.
- The southern tip of the continent is called *Tierra del Fuego,* or Land of Fire. The city found here, Ushuaia, is farther south than any other city in the world. Mark it with a red X.

Map of South America

Rain Forest Life

South America is home to the largest tropical rain forest in the world. Lying within the basin of the Amazon River, it covers much of the country of Brazil, and extends across the continent into parts of Bolivia, Peru, Ecuador, Colombia, Venezuela, Guyana, Suriname, and French Guiana. Most of the rain forest lies just south of the equator, and it rains nearly every day of the year.

The South American rain forest is home to many unusual plants, including ferns, orchids, and many varieties of trees. There are also unusual animals, such as the sloth, the spider monkey, and the red-eyed tree frog. Among the many varieties of rain forest birds are the parrots, which include the brightly colored Scarlet Macaw. The Scarlet Macaw is bright red with blue and yellow wing feathers. It lives in the tops of rain forest trees, and eats fruit, seeds, and nuts. Scarlet Macaws have become an endangered species because their habitat is disappearing and because many of them are captured to be kept as pets.

Project

Make a three-dimensional model of a Scarlet Macaw.

Materials

- Pattern page, following
- Photo reference of a Scarlet Macaw
- Crayons or markers
- Construction paper: red and a variety of other colors
- Paper or plastic cup
- Scissors
- Glue
- Yarn for hanging

Directions

1. Reproduce the pattern page.
2. Using the photo reference, color the head and wing pieces to match the coloring of the Scarlet Macaw. Cut out the pieces.
3. Cover the cup with construction paper. Glue the paper into place.
4. Cut strips of construction paper in other colors to create tail feathers. Glue into place.
5. Glue head and wing pieces to the cup.
6. Attach a length of yarn to the cup and suspend the Scarlet Macaw from the classroom ceiling.

Scarlet Macaw Pattern

Exploring South America

South America is a continent of many cultures, land types, and climates. The northern part of the land lies within the tropics, and is home to the world's biggest rain forest. The southernmost tip of the continent is only about 600 miles (966 km) from Antarctica.

Unscramble the words below to identify some of the interesting land features and other facts about South America. Write your answers in the space below each scrambled word. Use resource books or the Internet to help you find the answers.

1. This is the second-largest river in the world, after the Nile River, and is surrounded by rain forest.	Aaomnz ivrRe
2. This waterfall in Venezuela is the highest waterfall in the world.	gAenl alFls
3. These are the central plains of Argentina, home to South American cowboys, called *gauchos*.	papsam
4. This region at the southern tip of Argentina is an area of rocky islands with glaciers, mountains, and valleys.	errTia eld uFoeg
5. This western mountain range extends the entire length of the continent, and is the longest mountain range in the world.	ndAse
6. This is the largest country on the South American continent.	Brlzia
7. These animals are found in the Andes mountains, and are used as beasts of burden, and also for their wool.	lamlas
8. This lake, which is on the border between Peru and Bolivia, is the world's highest navigable lake.	Lkae Ttiaacci
9. This South American country shares a border with Panama, which is in Central America.	Cooabiml
10. This is the long, narrow country that stretches down the Pacific coast of South America.	hilCe
11. This mountain, found in Argentina, is the highest point on the South American continent.	Mntuo Acacgnuoa
12. These islands, off the coast of Ecuador, are known for their unusual wildlife.	Glapaasgo slInads

Answer Key: 1. Amazon River 2. Angel Falls 3. pampas 4. Tierra del Fuego 5. Andes 6. Brazil 7. llamas 8. Lake Titicaca 9. Colombia 10. Chile 11. Mount Aconcagua 12. Galapagos Islands

Europe

Europe is the sixth-largest of the world's continents. Europe is bordered by the Arctic and Atlantic Oceans in the north and west, and the Mediterranean Sea in the south. The eastern border is shared with the continent of Asia. Europe is the only continent in the world without a desert, its land being covered with mountains, forests, rivers, and fields. There are three major mountain ranges: the Alps, the Pyrenees, and the Carpathian Mountains. There are more than 40 countries in Europe, each with its own languages and cultures.

The far northern edge of Europe has an arctic climate and is covered with arctic tundra. Iceland is a volcanic island nation in the North Atlantic. In Scandinavia, the land is covered with forest. Central and Eastern Europe lie on a great plain, with warm summers and cold winters. The United Kingdom and Ireland, northwestern islands, have a very mild climate, regulated by warm ocean currents. Lands such as Spain and Italy to the south have hot summers and mild winters.

Project
Create a map showing some of the main geographic features of Europe.

Materials
- Atlas, encyclopedia, or topographical map of Europe
- Map page, following
- Pen or pencil
- Colored pencils

Directions
Using the atlas, encyclopedia, or map as a guide, find the locations of the following geographic features and mark them on the map page. Be sure to label the features as you add them.

- Color and label these European countries:
  - Austria
  - Belgium
  - Denmark
  - Finland
  - France
  - Germany
  - Greece
  - Iceland
  - Ireland
  - Italy
  - Luxembourg
  - Netherlands
  - Norway
  - Poland
  - Portugal
  - Russian Federation
  - Spain
  - Sweden
  - Switzerland
  - United Kingdom

- Draw Europe's longest river, the Volga, in blue.
- Draw mountain peaks to show the location of these mountains ranges: the Alps, the Pyrenees, and the Carpathian Mountains.
- Label the Atlantic Ocean and the Mediterranean Sea.

- Find these European capital cities and mark each location with a star: London, Paris, Rome, Madrid, Berlin.
- Locate and label these waterways: the English Channel, the Baltic Sea, the Black Sea, the Adriatic Sea.

Map of Europe

European Landmarks

Europe is a continent of many beautiful landscapes and fascinating cities. But if you ask people to name one symbol that represents Europe, many will give the same answer: the Eiffel Tower in Paris, France. This iron tower rises to a height of 1,062 feet (324 m), including the flagpole. Opened on March 31, 1889, it was designed by structural engineer Gustave Eiffel for the Universal Exhibition in celebration of the French Revolution. The open, lattice-like appearance of the tower made it a very unusual structure for the time.

For many years, the Eiffel Tower was the highest structure in the world. Tourists can ride elevators to one of three observation platforms or to the top, or they can ascend the tower using the 1,665 steps. The tower was used during World War I as an important military observation station, and is still used as a television and radio transmitting tower today.

Project

Create a yarn picture of the Eiffel Tower.

Materials

- Photo resources of the Eiffel Tower
- White art paper or construction paper
- Thin, black yarn
- Glue
- Pencil
- Colored pencils or crayons
- Black construction paper

Directions

1. Using the photos for guidance, use a pencil to sketch the Eiffel Tower. Be sure to show the lattice-like, open design of the iron structure.
2. Carefully glue strands of yarn over the pencil lines to represent the iron pieces that make up the building. Let dry.
3. Use colored pencils to draw a Paris street scene at the base of the tower.
4. Create a frame for your picture by gluing it to a piece of black construction paper.

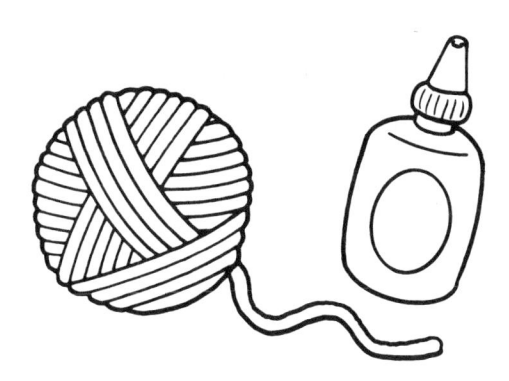

Countries of Europe

Europe is not a large continent, but it is home to over 40 countries. Even within individual countries, there are sometimes a variety of languages spoken and different cultures represented. The northernmost region of Europe is called Scandinavia. The countries there include Norway, Sweden, and Denmark. Most of the land is covered with forests, and the climate is very cool all year long. Western Europe includes the British Isles, as well as large nations such as Spain, Italy, Germany, and France. There are many types of land here, including the Alps, the highest mountains in Europe. There is a mixture of farmland and cities all across this area. The weather in places like Spain and Italy is warm, making them popular vacation areas. Eastern Europe includes countries such as Poland, Hungary, and Romania. Here the winters are cold and the summers warm. In Eastern Europe as well as in much of Western Europe, there are big cities that contain factories and apartment buildings.

Project
- Create a "Europe Fact Board" with information about individual countries in Europe.
- Complete a word search, looking for the names of European countries.

Materials
- Word search page, following
- Pen or pencil
- Research materials
- Index cards

Directions
1. Select a European country and use research materials to learn about it.
2. Find at least three interesting facts about your country. Write each fact on an index card.
3. Select your favorite fact card. After writing the name of the country at the top of the card, post it on a "Europe Fact Board" to create a classroom display.
4. Complete the word search on the following page. The names of the countries to be found are in the box at the right. After completing the word search, find each of the countries on a map of Europe.

Albania	Italy
Austria	Hungary
Belarus	Netherlands
Belgium	Norway
Denmark	Poland
Finland	Portugal
France	Spain
Germany	Sweden
Greece	Switzerland
Ireland	United Kingdom

Countries of Europe

S	W	I	T	Z	E	R	L	A	N	D	S	W	E	D	E	N	S
W	Q	E	R	T	Y	U	I	O	P	D	A	S	D	F	G	H	J
K	L	Z	Y	C	X	V	B	N	M	E	Z	X	B	C	V	B	H
N	M	Q	L	W	R	T	Y	U	I	N	O	E	E	P	A	S	U
D	F	Y	A	W	R	O	N	G	H	M	L	J	L	K	L	Z	N
X	C	V	T	B	N	M	Q	W	E	A	R	T	G	Y	U	Q	G
W	E	F	I	N	L	A	N	D	R	R	R	T	I	Y	U	I	A
O	P	A	S	D	F	G	H	U	J	K	K	L	U	Z	X	C	R
A	V	B	N	M	Q	W	S	E	R	T	Y	U	M	I	F	O	Y
U	N	I	T	E	D	K	I	N	G	D	O	M	P	A	R	S	D
S	F	G	H	J	K	L	Z	X	C	V	B	N	M	Q	A	W	E
T	R	T	Y	U	I	O	P	A	P	S	D	F	G	H	N	J	K
R	L	Z	X	C	V	B	N	M	O	Q	G	R	E	E	C	E	W
I	R	E	L	A	N	D	Y	U	R	I	E	O	A	P	E	A	S
A	E	R	T	Y	N	D	F	G	T	H	R	J	L	K	L	Z	X
C	V	B	N	A	M	S	W	E	U	R	M	T	B	Y	U	I	O
P	A	S	L	D	F	P	G	H	G	J	A	K	A	L	C	V	B
N	M	O	Q	W	E	A	R	T	A	Y	N	U	N	I	O	P	A
S	P	D	F	G	H	I	J	K	L	L	Y	Z	I	X	C	V	B
N	H	Y	B	G	T	N	E	T	H	E	R	L	A	N	D	S	J
U	J	M	I	K	O	V	F	R	C	D	E	X	S	W	A	Q	N

Continents Activity Book 17 ©Edupress, Inc.™ EP093

Africa

Africa is the second-largest continent. It is surrounded by the Atlantic Ocean, the Indian Ocean, the Mediterranean Sea, and the Red Sea. The continent is divided into 54 countries. Its southernmost point is Cape Agulhas, in South Africa.

There are mountain ranges in the extreme northwest and southeast of Africa, but most of the continent consists of vast flat lands or plateaus divided by many rivers and streams. Deserts cover nearly forty percent of the continent. The Sahara, in northern Africa, is the largest desert in the world. Savannas, or grasslands, cover another forty percent of the continent. The savannas extend from south of the Sahara to the Congo Basin. The remainder of the continent consists of rain forests, located in the Congo Basin and in parts of western Africa and Madagascar.

Project
Create a map showing some of the main geographic features of Africa.

Materials
- Atlas, encyclopedia, or topographical map of Africa
- Map page, following
- Pen or pencil
- Colored pencils

Directions
Using the atlas, encyclopedia, or map as a guide, find the location of the following geographic features and mark them on the map page. Be sure to label the features as you add them.

- Color and label these African countries:
    - Algeria
    - Angola
    - Botswana
    - Chad
    - Egypt
    - Ethiopia
    - Kenya
    - Libya
    - Madagascar
    - Morocco
    - South Africa
    - Sudan
    - Tanzanie
- Identify the Atlantic Ocean, the Indian Ocean, the Mediterranean Sea, and the Red Sea.
- Find the Nile River and trace it in blue.
- Locate the Sahara Desert and color it yellow. It is the largest desert in the world.
- There are two other large deserts on the continent—the Kalahari and the Namib. Find them and color them yellow.
- Determine which parts of the continent are rain forests. Color them green.
- Africa's largest rain forest is in the Congo River Basin. Find the Congo River and trace it in blue.
- The highest mountain on the African continent is Mt. Kilimanjaro in Tanzania. Find its location and mark it with a star.
- Lake Victoria is the largest lake in Africa. Locate it and color it blue.

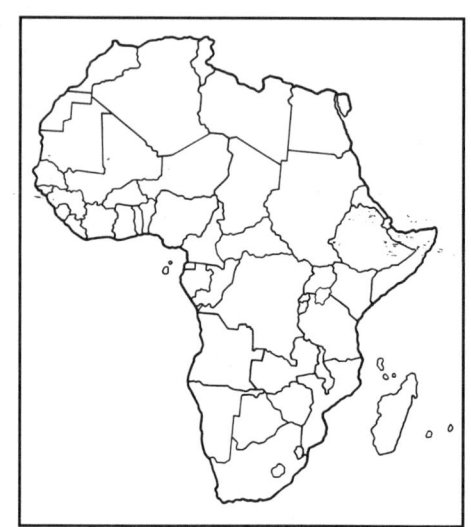

Map of Africa

Countries of Africa

There are 54 countries in the continent of Africa. Some of the countries, like Madagascar, are islands. Egypt is a very old country. Other countries have only become nationalized in the last 100 years. Sudan is the largest African country; Seychelles is the smallest.

Because the continent of Africa is so large, the climate and living conditions can be very different from place to place. North Africa is mostly desert. West Africa is rain forest and farmland, where farmers grow coffee, cocoa, and rubber trees. Central Africa is dominated by the Congo River Basin, which is a heavy tropical rain forest. Transportation in Central Africa is almost all by river boat. East Africa is a land of mountains and grasslands, home to many of Africa's wild animals. South Africa is rich in diamonds, gold, iron, and copper. There are more diamond and gold mines here than in any other place in the world.

Project

- Create an "Africa Fact Board" with information about individual countries in Africa.
- Complete a word search, looking for the names of African countries.

Materials

- Word search page, following
- Pen or pencil
- Research materials
- Index cards

Directions

1. Select an African country and use resource materials to learn about it.
2. Find at least three interesting facts about your country. Write each fact on an index card.
3. Select your favorite fact card. After writing the name of the country at the top of the card, post it on an "Africa Fact Board" to create a classroom display.
4. Complete the word search on the following page. The names of the countries to be found are in the box to the right. After completing the word search, find each of the countries on a map of Africa.

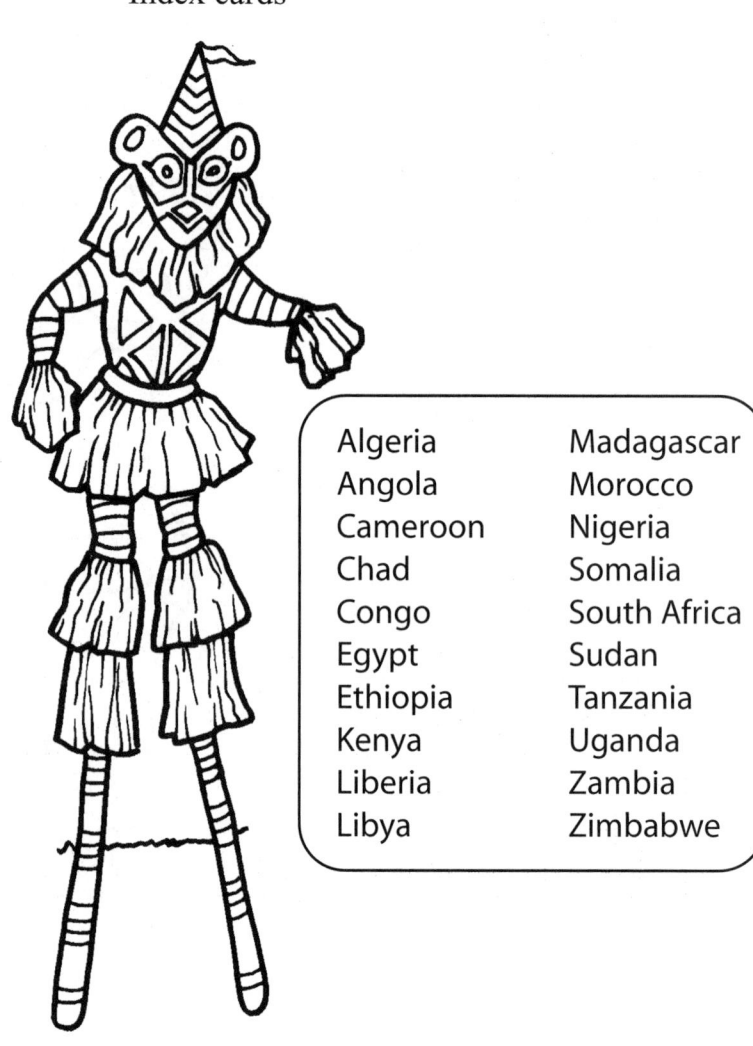

Algeria	Madagascar
Angola	Morocco
Cameroon	Nigeria
Chad	Somalia
Congo	South Africa
Egypt	Sudan
Ethiopia	Tanzania
Kenya	Uganda
Liberia	Zambia
Libya	Zimbabwe

Countries of Africa

U	G	A	N	D	A	E	S	M	G	A	I	T	E	Y	K	N	O
P	B	I	O	A	I	B	M	A	Z	W	S	U	D	A	N	H	A
A	Z	A	M	H	B	E	O	R	O	E	F	R	A	Y	I	S	N
Y	M	A	D	A	G	A	S	C	A	R	C	O	O	N	U	A	T
A	I	L	E	T	I	A	L	P	C	A	L	F	A	E	W	E	M
I	A	G	Y	C	S	A	C	I	T	L	A	G	R	K	I	C	L
O	A	E	P	O	B	G	Y	R	B	G	T	P	Y	G	E	C	P
Z	R	R	A	N	O	A	H	T	M	E	A	T	S	D	B	R	A
S	L	I	M	G	A	C	I	A	D	B	R	O	E	P	A	L	Z
O	T	A	G	O	I	M	N	N	O	A	T	I	W	S	D	H	A
M	Z	N	M	H	E	A	E	Z	C	N	A	T	A	O	T	B	C
A	H	R	E	G	H	C	G	A	M	R	I	O	R	U	R	I	A
L	A	R	T	G	R	I	M	N	A	I	R	N	A	T	S	P	M
I	A	D	A	P	T	N	A	I	G	T	E	O	R	H	I	E	E
A	N	G	O	L	A	O	M	A	G	I	G	C	E	A	E	I	R
P	L	D	E	G	U	I	D	H	N	E	I	C	L	F	R	L	O
E	W	B	A	B	M	I	Z	B	R	A	N	O	B	R	E	I	O
I	A	E	L	A	O	F	B	E	N	O	Y	R	A	I	M	A	N
B	R	E	T	H	I	O	P	I	A	M	A	O	H	C	H	G	O
K	M	A	I	M	T	A	M	R	E	A	D	M	O	A	I	M	S
N	T	L	B	I	L	I	B	Y	A	E	I	T	S	T	A	B	D

The Sahara Desert

The Sahara Desert, in the northern part of Africa, is the largest desert in the world. It covers an area of 3.5 million square miles (9 million sq km), and stretches from the Atlantic Ocean to the Red Sea. The highest recorded temperature in the world was in the Sahara, where it reached 136.4° Fahrenheit (58°C). Like most deserts, nearly all of the Sahara Desert is covered with sand.

The hot, dry climate of the Sahara makes it unwelcoming to many living things. There are very few plants, although palm trees and shrubs can live in some areas. Animals that live in the Sahara have adapted to the hot weather. Many are nocturnal, sleeping in a cool den or cave during the day and hunting at night. Birds fly high in the air to escape the hot temperatures close to the desert floor. One of the few mammals to live in the Sahara is the camel. It adapts by drinking large amounts of water and storing it for long periods of time in either one or two humps on its back. The people who live in the Sahara live mainly along the edges of the desert. They dress in clothes that help them stay cool and protect them from the hot sun.

Project

Paint a picture of the Sahara Desert.

Materials

- Art paper
- Paint and paintbrush
- Desert photos, including pictures of the Sahara Desert

Directions

1. Look at the photos of deserts, taking note of the things you observe.
2. Using the information provided above as well as the photos, paint a picture of how you think the Sahara Desert looks. Be sure to include plants and animals in your picture.
3. Label all of the things in your painting.
4. Compare your picture to the photos you viewed earlier. Does your desert picture look like the photos?

Asia

Asia is the biggest continent, taking up one-third of the world's land surface. It extends all the way from the cold Arctic Ocean to the tropical islands in the East Indies. There are close to 50 countries in Asia. On the west it is bordered by the Ural Mountains, the Black Sea, and the Mediterranean Sea. To the east is the Pacific Ocean. The continent is divided from east to west by the Himalaya Mountains, which include the world's highest mountain peaks.

Some Asian countries have huge populations. Areas such as Japan and southern China are very crowded. Other areas remain wilderness and have few inhabitants. The northernmost part of Asia is Siberia, part of the Asian portion of the Russian Federation. The winters in Siberia are bitterly cold. In contrast, the lands in Southwest Asia (the regions near Turkey and Saudi Arabia, adjacent to Africa) have much warmer climates and desert landscapes.

Project

Create a map showing some of the main geographic features of Asia.

Materials

- Atlas, encyclopedia, or topographical map of Asia
- Map page, following
- Pen or pencil
- Colored pencils

Directions

Using the atlas, encyclopedia, or map as a guide, find the location of the following geographic features and mark them on the map page. Be sure to label the features as you add them.

- Color and label these Asian countries:
    Afghanistan   Armenia
    China   India
    Indonesia   Iran
    Iraq   Japan
    Mongolia   North Korea
    Pakistan   Philippines
    Russian Federation Saudi Arabia
    Singapore   South Korea
    Taiwan   Thailand
    Turkey   Vietnam

- Draw mountain peaks to show the location of the Himalaya Mountains. Draw a red star to show the location of Mt. Everest, the tallest mountain in the world.

- Identify the Arctic Ocean, the Pacific Ocean, and the Indian Ocean.

- There is a chain of mountains that runs the length of the country of Japan. Draw a red star to show the location of Mt. Fuji, Japan's most famous mountain.

- The longest river in Asia is the Chang (or Yangtze) River in China. Draw it in blue.

- There are two important rivers in Iraq: the Tigris and the Euphrates. Draw them in blue.

- The Gobi Desert, one of the largest deserts of the world, is in Mongolia. Find its location and color it yellow.

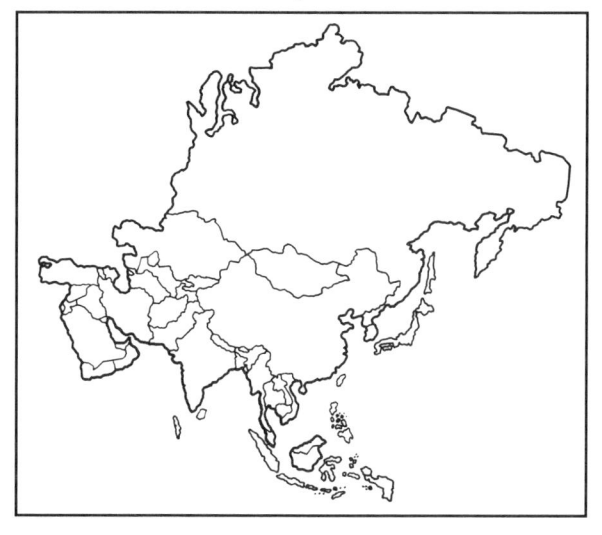

Map of Asia

Find Out About Asia

Asia is a continent of diverse people and countries, from Asian Russia in the far north to Indonesia in the south.

Unscramble the words below to identify some of the interesting land features and other facts about Asia. Write your answers below the scrambled words. Use resource books or the Internet to help you find the answers.

1. Asia contains the two most populous countries in the world. Name them.	anCih ndIia
2. This area is the part of Asia that lies between the Mediterranean Sea and India.	deMldi sEta
3. This mountain lies on the border of Tibet and Nepal, and is the highest mountain in the world.	ouMtn veEtrse
4. Japan is an Asian nation made up of islands. Two of the main islands are called Shikoku and Kyushu. Name the other two.	hsHuno Hakodiko
5. This Chinese river is the third-longest river in the world.	Yzegtan
6. This huge desert covers much of Mongolia.	obGi eseDrt
7. This area of Asia contains India, Pakistan, Nepal, Bhutan, Bangladesh, and Sri Lanka.	Iandin sconetintnub
8. This mountain range lies between India and Tibet, and includes the highest mountain in the world.	mayaalHi Mtaunsino
9. Most of this river runs through India, although it empties into the Indian Ocean in Bangladesh.	negsGa Reriv
10. These two countries lie on the peninsula between China and Japan.	rtNho oreKa Sutho Keaor
11. This Asian country lies between the Black Sea and the Mediterranean Sea.	Teyrku
12. This Asian country is the largest archipelago, or group of islands, in the world.	doInsinae

Answer Key: 1. *China, India* 2. *Middle East* 3. *Mount Everest* 4. *Honshu, Hokkaido* 5. *Yangtze* 6. *Gobi Desert* 7. *Indian subcontinent* 8. *Himalaya Mountains* 9. *Ganges River* 10. *North Korea, South Korea* 11. *Turkey* 12. *Indonesia*

Mount Fuji

The variety of landscapes found on the continent of Asia is huge, from the forests of northern Russia to the vast deserts of Mongolia and Saudi Arabia. The continent is divided by mountain ranges that contain some of the highest mountains in the world. There are also large tropical areas, such as in the islands of Indonesia.

One of the most famous natural landmarks in Asia is Mount Fuji, the highest mountain in Japan. Mount Fuji is an inactive volcano, and it rises 12,388 feet (3,776 m) above the island of Honshu. The top of the mountain is often covered in mist so thick that the people living on the island cannot see it. The people of Japan have long considered Mount Fuji a sacred mountain, and each year more than 50,000 pilgrims climb to the summit.

Project

Paint a picture of Mount Fuji, a very popular theme for Japanese art.

Materials

- White paper
- Tempera paint
- Paintbrushes
- Cotton balls
- Glue
- Hole punch
- Yarn
- Colored construction paper

Directions

1. Use paint and paintbrushes to paint a picture of Mount Fuji. Use black paint to copy the Japanese character for "mountain" as seen in the illustration. Let the painting dry.

2. Pull apart a cotton ball to look like wisps of cloud. Glue the cotton to the painting at the mountain crest.

3. Glue the painting onto a larger sheet of colored construction paper.

4. Use a hole punch to make two holes at the top of the construction paper. Make a hanger from yarn.

Australia

Australia is the only country that is also a continent. It ranks as the sixth-largest country in the world in area, but it is the smallest continent. Australia is often grouped together with Papua New Guinea, New Zealand, and the many islands of the South Pacific. Together they are called Australasia or Oceania. Australia lies between the Indian and South Pacific Oceans, and is often referred to as being "down under," because it lies entirely in the Southern Hemisphere.

Australia is divided into six states: New South Wales, Queensland, South Australia, Tasmania, Victoria, and Western Australia. There are also two mainland territories: the Australian Capital Territory and the Northern Territory. Only a few areas along or near the coasts receive enough rain to support a large population.

The interior of the Australian continent, dry and desert-like, is known as the outback, home to miners and farmers. Northern Queensland is a land of tropical rain forests, and the Pacific coast is edged by the world's biggest coral reef, the Great Barrier Reef. New South Wales and Victoria are covered with eucalyptus forests and grasslands where huge flocks of sheep are raised.

Project
Create a map showing some of the main geographic features of Australia.

Materials
- Atlas, encyclopedia, or topographical map of Australia
- Map page, following
- Pen or pencil
- Colored pencils

Directions
Using the atlas, encyclopedia, or map as a guide, find the locations of the following geographic features and mark them on the map page.

- Label each of the six states and two territories.
- Mark the state capitals with a dot and label them: Perth, Melbourne, Sydney, Brisbane, Adelaide, and Hobart. Mark and label the capital city of the Northern Territory, Darwin.
- Locate and label the national capital, Canberra. Identify it with a star.
- Find and label these bodies of water: Indian Ocean, Pacific Ocean, and Southern Ocean.
- Locate the central area known as the outback. Color it light brown.

- The biggest lake on the continent is Lake Eyre. Locate it and color it blue.
- The Great Barrier Reef can be found along the eastern shore of Queensland. Find its location and label it.
- Northern Queensland is covered with tropical rain forests. Color this area green.
- The Great Dividing Range is a mountain range that runs along the eastern coast of the continent. Use a black colored pencil to draw the mountains and label the mountain range.
- Find and label the Murray-Darling River. Trace it in blue.

Map of Australia

Around Australia

Australia is a land of unusual land features and unique animals. Its population is diverse, and its customs are influenced by its aboriginal natives as well as Europeans who settled the land.

Unscramble the words below to identify some of the interesting land features and other facts about Australia. Write your answers below the scrambled words. Use resource books or the Internet to help you find the answers.

1. These people were the original inhabitants of Australia, and have a rich heritage of art and folklore.	**bgionesirA**
2. This city is the capital of Australia.	**anbeaCrr**
3. Large ranches in the Australian interior where ranchers raise sheep are called ____. (two words)	**seeph stinstoa**
4. The largest bird in Australia is the _____.	**coarywssa**
5. This is the Australian state that is an island.	**miaanTsa**
6. The capital of New South Wales, this city is famous for its beautiful, unique opera house.	**ndSeyy**
7. This large coral reef lies off the eastern coast of Queensland.	**atGer Berriar feeR**
8. Although it looks like a bear, this animal is a marsupial. It lives in eucalyptus trees and eats eucalyptus leaves.	**alaok**
9. The Aborigines consider this huge, red rock in the Australian outback a sacred spot.	**ersAy Rcko**
10. The dusty interior of the Australian continent is known as the ____.	**ubatock**
11. The principal river in Australia is the ____. (two words)	**Mryrau-arlngDi**
12. Because it is so far south, Australia is sometimes known as "The Land ____." (two words)	**Dwno drneU**

Answer Key:
1. Aborigines 2. Canberra 3. Sheep stations 4. Cassowary 5. Tasmania 6. Sydney 7. Great Barrier Reef 8. Koala 9. Ayers Rock 10. Outback 11. Murray-Darling 12. Down Under

The Great Barrier Reef

One of Australia's most famous natural features is the Great Barrier Reef, a coral reef that stretches for 1,250 miles (2,010 km) along the northeast coast of Queensland. It is the largest coral reef in the world. The climate, the sandy beaches, and a great variety of animal life make the Great Barrier Reef a popular destination for tourists.

Coral reefs are formed by tiny creatures called *polyps*. When polyps die, their outer skeletons are left behind. This is known as *coral*. Other polyps become attached to these skeletons, and when they die, their skeletons add to the reef. This is how a coral reef grows.

The Great Barrier Reef is composed of about 400 species of coral in many shapes and colors. It is also home to about 1,500 different species of fish and other creatures, including sponges, starfish, crabs, sea urchins, sea cucumbers, turtles, and clams. It is the world's largest marine park.

Project

Create a sea life dictionary featuring animals found in the Great Barrier Reef.

Materials

- Art paper
- Photo resources for sea life native to the Great Barrier Reef (may include magazines or Internet access)
- Crayons, markers, or colored pencils
- Stapler

Directions

1. Use resources to find out about some of the species of sea life found in and around the Great Barrier Reef: sponges, starfish, crabs, sea urchins, sea cucumbers, turtles, clams, and coral.

2. On individual pages, draw a picture of one form of sea life and write a descriptive passage about it.

3. Staple the pages together to create a Great Barrier Reef sea life dictionary, or post the pictures on a Great Barrier Reef Sea Life bulletin board.

Antarctica

Antarctica is the fifth-largest continent. It sits at the bottom of the world, surrounded by the Atlantic, Pacific, and Indian Oceans. The South Pole lies near the center of the continent. Nearly all of the continent is covered in a layer of ice that constantly moves, in the form of glaciers inching toward the ocean. Chunks of ice break off the continent and float as icebergs. Antarctica is the coldest continent, and heavy winds blow most of the time. It can be classified as a desert, because less than two inches (5 cm) of rain or snow falls every year. There are many mountains there, and several active volcanoes.

There is very little vegetation on Antarctica, although there are animals that live on the ice and in the oceans. It is too cold for people to make permanant settlements, but scientists visit and stay there long enough to conduct research.

Project

Create a map showing some of the main geographic features of Antarctica.

Materials

- Atlas, encyclopedia, or topographical map of Antarctica
- Map page, following
- Pen or pencil
- Colored pencils

Directions

Using the atlas, encyclopedia, or map as a guide, find the location of the following geographic features and mark them on the map page. Be sure to label the features as you add them.

- Find and label the Atlantic Ocean, Indian Ocean, and Pacific Ocean.
- Identify the Weddell Sea, the Amundsen Sea, and the Ross Sea.
- Big chunks of ice called icebergs break away from Antarctica and float out to sea. Draw some icebergs in the water around the continent. Color them blue.
- The south pole is the southernmost point on the Earth. It is marked by flags and a pole. Find the South Pole and mark it with a red and white pole.
- An ice shelf is a thick sheet of ice that sticks out beyond the land into the sea. Locate the Ronne Ice Shelf and the Ross Ice Shelf on Antarctica and color them purple.
- Draw blue mountain peaks to show the location of the Transantarctic Mountains.
- An imaginary line called the Antarctic Circle surrounds the continent. Draw dotted lines to show the circle.
- Mt. Erebus is the southernmost active volcano in the world. Find its location and mark it with a red X.

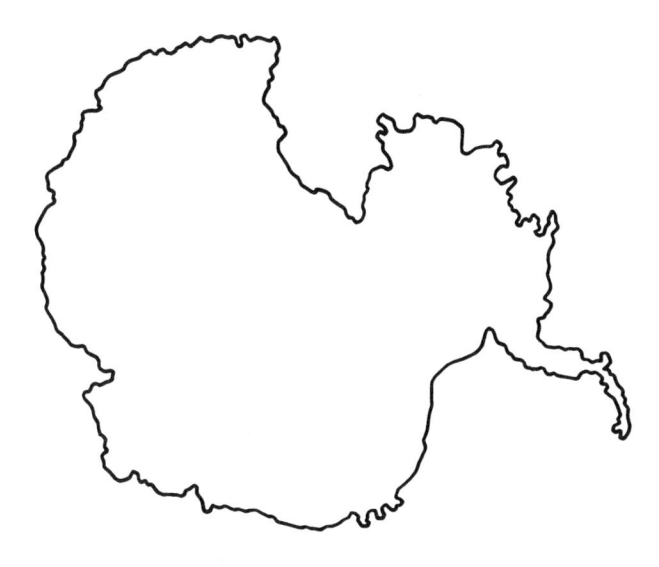

Map of Antarctica

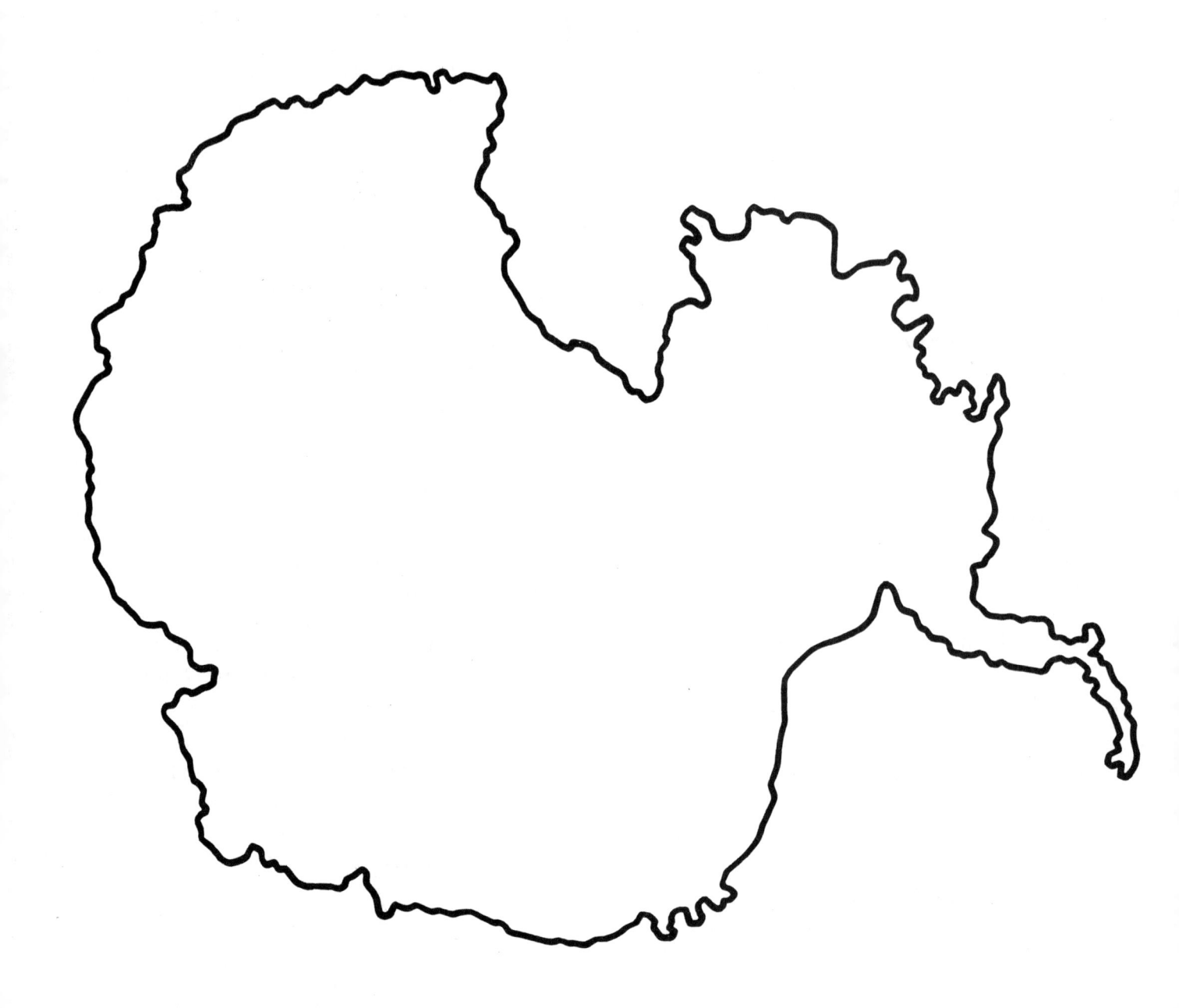

Amazing Antarctica

Antarctica is a fascinating continent, very different from the other six continents of the world. Complete the crossword puzzle below to learn some facts about the ice-covered land mass. You will find the words you need to complete the puzzle in the box below. Use resource books or the Internet to help you find the answers.

Across

5. Antarctica is surrounded by an imaginary circle called the _____. (2 words)
7. The most famous volcano on Antarctica is _____. (2 words)
8. Much of Antarctica is covered with frozen rivers of ice called _____.
9. Big chunks of ice that break away from Antarctica and float in the sea are called _____.
10. At the center of Antarctica is the _____. (2 words)

Down

1. The layer of ice over the land of Antarctica is called an _____. (2 words)
2. Antarctica is surrounded by three oceans: the Atlantic, the Pacific, and the _____ Oceans.
3. Two of the species of penguins that live in Antarctica are Emperor penguins and _____ penguins.
4. The highest mountain in Antarctica is _____. (2 words)
6. Antarctica is the home of flat-footed, flightless birds called _____.

Adelie
Antarctic circle
glaciers
icebergs
ice cap
Indian
Mt. Erebus
penguins
South Pole
Vinson Massif

Antarctic Food Web

Because of the harsh climate, the continent of Antarctica supports very little plant and animal life. However, the water around the continent supports a complex food web, which is a system of plants and animals that depend on each other for food.

The Southern Ocean is rich in nutrients, and tiny plants and animals known as plankton thrive in the cold water. Krill, a shrimplike animal that grows no bigger that 2½ inches (6.4 cm), eats the tiny plants. Krill is the most important food source in the Antarctic food chain. Sea birds, baleen whales, and seals eat krill. Someday, krill may be used as food in many countries.

Project

Learn about how plants and animals depend on each other by making a paper chain that illustrates one part of the Antarctic food chain.

Materials

- Food chain patterns, below
- Crayons or colored pencils
- Resource materials
- Paper for additional food chain
- Scissors
- Tape

Directions

1. Reproduce the pattern page, below.
2. Color and cut out the food chain "links." Form a chain by linking a food source to the animal that eats it.
3. There are many other food chains in the Antarctic food web. For example, a baleen whale eats krill, which eats plankton. Use resource materials to research and make another chain that illustrates how other Antarctic plants and animals depend on each other.

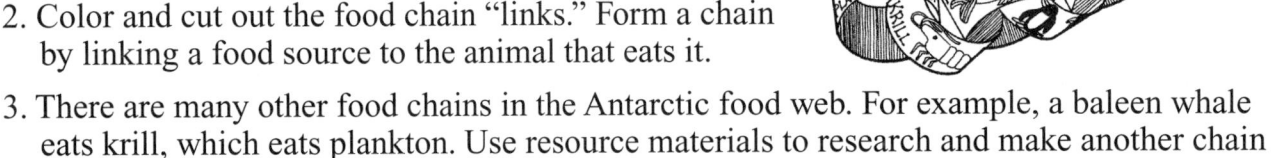

Toothed Whales		Eat Seals
Seals		Eat Penguins
Penguins		Eat Fish & Squid
Fish	Squid	Eat Krill
Krill		Eat Plankton
Plankton		

Cuisine from the Continents

Most countries of the world have favorite foods that are related to their culture, history, and celebrations. Learning about these foods can be an exciting, fun experience.

Use the recipes below to get a taste of foods from around the world.

Africa

Jollof Rice

Jollof rice comes from West Africa. "Jollof" means the rice is cooked in the dish rather than separately.

Ingredients

- 1 chicken, cut up
- 2 small cans tomatoes
- 2 cups (500 ml) water
- Salt & pepper to taste
- 1 cup (250 ml) uncooked rice
- ¼ tsp. (1 ml) each, cinnamon and ground red pepper
- 3 cups (750 ml) coarsely shredded cabbage
- 1 cup (250 ml) sliced green beans
- 2 onions, sliced

Directions

1. Combine chicken, tomatoes with liquid, water, salt, and pepper in a large pan. Heat to boiling, then reduce heat.
2. Cover and simmer 30 minutes.
3. Remove chicken. Stir in rice, cinnamon, and red pepper. Add remaining ingredients and return chicken to the pan.
4. Reduce heat, cover and simmer until chicken and rice are done, 20 to 30 minutes.

Serves 4-6

Australia

Anzacs

Anzacs are a popular dessert cookie in Australia. Anzac Day is a national holiday celebrated in remembrance of the Battle of Gallipolli in 1915.

Ingredients

- 1 cup (250 ml) butter, softened
- 1 cup (250 ml) flour
- 1 cup (250 ml) rolled oats
- 2 Tbsp. (30 ml) maple syrup
- 1 tsp. (5 ml) baking powder
- 1 cup (250 ml) flaked, unsweetened coconut
- 1 cup (250 ml) sugar

Directions

1. Cream butter and maple syrup in a large bowl. Add remaining ingredients.
2. Roll into small balls and place well apart on greased cookie sheets.
3. Bake at 350°F (180°C) for about 15 minutes.
4. Check for doneness. Cool on cookie sheets before removing.

Europe

Smorrebord (Open Sandwich)

Smorrebord are very popular in Scandinavia. The basic ingredients are a piece of buttered bread with meat, cheese, and other toppings.

Ingredients
Bread
Butter
An assortment of sliced meats such as turkey or roast beef
An assortment of sliced cheeses
Chopped pickles, eggs, tomatoes, and other toppings as desired

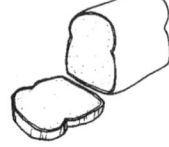

Directions
1. Butter a slice of bread on one side and lay it on a plate.
2. Use other ingredients to create an open-face sandwich.

Asia

Chicken Egg Soup

This is a Japanese soup. A similar soup, called Egg Flower Soup, is served in China.

Ingredients
6 cups (1.4 l) canned chicken broth
1 Tbsp. (15 ml) soy sauce
3 eggs, beaten
1 stalk celery, minced

Directions
1. Bring chicken broth and soy sauce to a boil. Turn off heat.
2. Slowly pour egg into the soup. The egg will form shreds.
3. Sprinkle celery on soup after it is placed in serving bowls.

Serves 8

Greek Salad

This salad should be served at room temperature.

Ingredients
6 Tbsp. (88 ml) olive oil
Juice of 1 lemon
3 cloves garlic, minced
1 head of Romaine lettuce, torn into pieces
3 tomatoes, cut into wedges
12 Greek black olives, in oil
1 onion, thinly sliced
¼ cup (59 ml) flat leaf parsley
½ cup (118 ml) feta cheese, crumbled

Directions
1. Put oil, lemon juice, and garlic in a small jar. Tighten lid and shake well.
2. Place remaining ingredients in a large bowl. Pour dressing over salad and toss gently.

Serves 6

Ch'ao-Hsueh Tou (Stir-Fried Snow Peas)

This stir-fried vegetable dish is from China.

Ingredients
½ lb. pork with some fat, finely chopped
1 lb. snow peas, washed, dried, and trimmed
½ cup (125 ml) chicken broth
4 Tbsp. (60 ml) water
1 Tbsp. (15 ml) cornstarch
1 Tbsp. (15 ml) soy sauce

Directions
1. Fry pork in a wok or large skillet until crisp and cooked through.
2. Stirring continually, add snow peas. Reduce heat to medium and cook for one minute.
3. Still stirring, add chicken broth and continue until snow peas are tender but still crisp.
4. Combine soy sauce and cornstarch. Add mixture to snow peas to thicken, mix well, and remove from heat.

Serves 4-6

North America

Nanaimo Bars

Many people consider this layered cookie the national dessert of Canada. Nanaimo Bars are named for the city of Nanaimo, in British Columbia.

Layer 1
- ½ cup (125 ml) butter
- ¼ cup (60 ml) sugar
- ½ cup (125 ml) cocoa
- 1 egg, beaten
- 1 tsp. (5 ml) vanilla
- 2 cups (472 ml) graham cracker crumbs
- ½ cup (125 ml) chopped walnuts
- ½ cup (125 ml) coconut

1. Melt sugar, butter, and cocoa in a double boiler. Let cool.
2. Add egg and vanilla. Add graham cracker crumbs, nuts, and coconut and mix well.
3. Cover the bottom of an 8-inch (20 cm) square baking pan with the mixture, packing it down well. Refrigerate.

Layer 2
- ¼ cup (60 ml) butter
- 2 cups (472 ml) powdered sugar
- 3 Tbsp. (45 ml) milk
- 2 Tbsp. (30 ml) instant vanilla pudding mix

1. Cream the butter and pudding mix.
2. Add milk and powdered sugar and beat until smooth.
3. Spread over the first layer and chill until firm.

Layer 3
- 3 squares semi-sweet chocolate
- 2 Tbsp. (30 ml) butter

1. Heat to melt. If mixture is too thick, add milk.
2. Pour over top of second layer. Refrigerate. Cut into 36 squares.

South America

Llapingachos (Peruvian Potato Cakes)

Potatoes are native to Peru. Over 200 varieties of potato were developed by the Incan and pre-Incan civilizations.

Ingredients
- 6 boiled potatoes, mashed, or 6 cups prepared instant mashed potatoes
- ¼ cup (60 ml) butter
- 2 onions, finely chopped
- 2 cups (472 ml) shredded Cheddar cheese
- Salt and pepper to taste
- Vegetable oil for frying

Directions
1. Put potatoes in a large bowl.
2. In a skillet over medium heat, cook onions in butter.
3. Add onions, pan scrapings, and cheese to the potatoes and mix well. Season with salt and pepper. Divide mixture into 12 patties.
4. Heat 2 tablespoons (30 ml) oil in skillet over medium-high heat. Fry patties until brown, about four minutes on each side. Add more oil as needed to prevent sticking.

Games Around the World

Although the seven continents are very different in terms of landforms and climate, the people who live on the continents are very much the same. Children everywhere love to play games. On these two pages, you'll find some games that children play in other parts of the world. Don't be too surprised if some of the games seem very familiar!

Yut

Korean families like to play this game on New Year's Day. It is played by two to four players.

You will need:

- Game board, drawn as in diagram at right
- Four game pieces (buttons or other markers) per player
- Four craft sticks, one side of each stick colored with a crayon or marker *(Yut sticks)*

How to play:

1. All markers start at the lower left-hand corner of the game board. *(Home)*
2. Player tosses the Yut sticks into the air, letting them land on the table.
3. Player counts the number of sticks that land with the colored side up and moves one of his or her markers the same number of spaces. If four sticks land with the uncolored side up, the player can move five spaces (because that is so difficult to do).
4. Play then goes to the next person. The first person with all four markers around the board wins.
5. Markers are moved around the board in a counterclockwise direction. If a marker lands on a corner spot, player may move diagonally across the square on his or her next play.

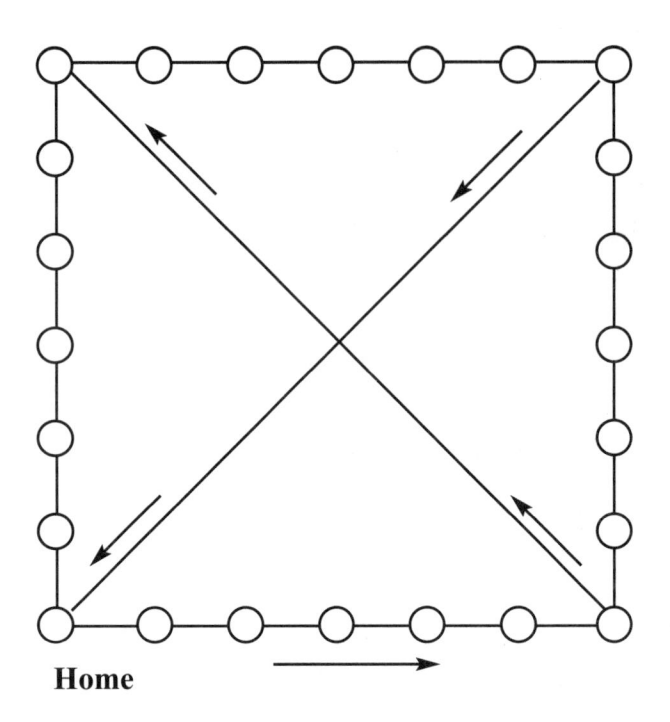

Cat and Mouse

This is a group game played by children in many African nations.

Players form a circle, but do not hold hands. One player, designated as the mouse, stands inside the circle. A second player, designated as the cat, stays outside the circle. The circle of players begins to move clockwise as the cat starts to chase the mouse, who runs in and out of the circle. A designated leader among the circle players can call for the circle to move the opposite direction. When the mouse is caught, new players are chosen.

Wall Ball
(Pelota pared)

Pelota pared is a group game. It is a favorite with children in Spain.

You will need:
- Tennis ball
- Chalk
- Wall or handball backboard

How to play:

1. Use the chalk to draw a line on the wall or backboard, about one yard (1 m) from the ground.
2. Assign a number to each player.
3. Player #1 bounces the ball on the ground and uses the palm of his hand to hit the ball against the wall or backboard. At the same time, he calls out the number of another player. The ball must strike the wall above the chalk line.
4. The player whose number is called must hit the ball before it bounces twice, calling out the number of another player.
5. A player is out if he or she fails to hit the ball before the second bounce, or if the ball strikes the wall below the chalk line.
6. The last player left after all others are eliminated is the winner.

Stop!

This easy game from Colombia is best with a large group of players.

You will need:
- Soft playground ball
- Large open space

How to play:

1. One player is designated the starter, and holds the ball. The remaining players gather around the starter.
2. The starter throws the ball straight up in the air, and at the same time calls out the name of another player. All of the players, including the starter, begin to run away at the moment the ball is thrown.
3. The player who is named by the starter must return and retrieve the ball. At the moment the player captures the ball, he or she shouts, "Stop!"
4. All of the players must stop when they hear the shouted command. The player with the ball then throws the ball gently at a nearby player, being sure to hit him or her below the waist.
5. If the player with the ball successfully hits another player, the one who is hit becomes the starter. If the throw is unsuccessful, the player who caught the ball is the starter.

Animals of the World

There are many factors that determine what kinds of animals live on each continent. These factors include temperature, the amount of rainfall the area receives, the type of vegetation, and the landforms found on the continent. All of these factors combine to create habitats, specific kinds of environments in which animals thrive. Each continent has more than one kind of habitat.

There are many animals that are common throughout the world. Some animals are not found all over the world, but only on a few continents. For example, elephants can be found in Asia and Africa. There are some animals that can only be found in a specific part of the world. The continent of Australia is home to several animals that are not found in the wild anywhere else.

Project #1

- Match animal description cards to the correct continent on the map.
- Do research to find facts about an unusual animal and draw a picture of that animal.

Materials

- Wall map of the world
- Animal description cards, page 41
- Tape, push pins, or thumbtacks
- Resource materials
- Pen or pencil
- Paper
- Crayons or other art media

Directions

1. Reproduce and cut apart the animal description cards on the following pages. Distribute them to individual students or student teams.

2. Challenge each student or student team to attach the animal description card to the continent on which the animal lives.

3. Use resource materials to find at least two more facts about each animal and write the facts on an index card.

4. Draw a picture of the animal to display with the fact cards.

5. Use the resource materials to make cards for other animals and add them to the map, as well.

Project #2

Create a student booklet about animals on the different continents.

Materials

- Animal coloring pages, pages 42-48
- Resource materials
- Stapler
- Pen or pencil
- Construction paper
- Crayons or other art media

Directions

1. Reproduce the coloring pages and color them.

2. Use resource materials to do research about the animals. Write at least one animal fact below each picture.

3. Staple the finished pages together with a construction paper cover.

Emperor Penguin The Emperor Penguin is the largest penguin. It lives in Antarctica, in groups called colonies. Penguins cannot fly, but they are good swimmers.	**Bengal Tiger** The Bengal tiger lives in different parts of Asia. It can live in a variety of habitats, including grasslands and rain forests.	**Giraffe** An adult giraffe is nearly 20 feet (6 m) tall. Giraffes feed on the leaves and buds at the tops of trees in the African savanna.
Polecat The polecat is the European relative of a skunk. It is a lone hunter that is active at night. Polecats mark their territories with a strong scent.	**Canada Goose** The Canada Goose is the common wild goose of North America. It makes a loud, honking sound. Canada Geese migrate south in the autumn.	**Giant Panda** The giant panda is found in the bamboo forests of China. It spends most of its time eating bamboo shoots and leaves. The giant panda is endangered.
Koala The Australian koala spends most of its life in eucalyptus trees, where its long toes and claws give it a good grip. The koala is a marsupial.	**Two-toed Sloth** The two-toed sloth is found in northern South America. It eats and sleeps hanging high in the trees.	**Alpine Ibex** The alpine ibex lives on the slopes of the Alps and other mountains in Europe. It will stand on its hind legs to reach the leaves on trees and shrubs.
Brazilian Tapir The Brazilian tapir is a rain forest animal that feeds on leaves, fruit, and water plants. It is a good swimmer.	**Platypus** The platypus is an aquatic, egg-laying mammal that lives in Australia. It has a beaver-like tail and a bill that looks like a duck's bill.	**Zebra** Herds of zebra roam the grasslands of Africa, eating grass, leaves, and bark. The zebra is related to the horse.
Weddell Seal The Weddell seal lives in the waters of Antarctica. Weddell seals communicate with each other by making sounds under water.	**Bison** Bison live in North America. At one time there were many bison on the prairie, but now there are only a few. Bison live in herds and feed on grass.	**Lion** Lions hunt in the grasslands of Africa and northwest India. Lions live in groups called prides. They eat animals such as antelopes and zebras.
Llama Llamas are related to camels. They are used as pack animals in the Andes mountains of Peru. They are also raised for their meat, wool, and hides.	**Dingo** The dingo is a wild dog that inhabits the dry plains and forests of Australia. Dingoes are nocturnal, and hunt both alone or in packs.	**Siamang** The siamang is an ape that lives in the forests of Asia. It is known for its loud barking call that can be heard up to two miles (3.2 km) away.

Animals of Africa

Giraffe

Lion

Zebra

Dromedary Camel

Animals of Antarctica

Emperor Penguin

Weddell Seal

Arctic Tern

Southern Right Whale

Animals of Asia

Giant Panda

Bengal Tiger

Siamang

Indian Elephant

Animals of Australia

Kangaroo

_____
_____
_____
_____

Duck-Billed Platypus

_____
_____
_____
_____

Cassowary

_____
_____
_____
_____

Koala

_____
_____
_____
_____

Continents Activity Book 45 ©Edupress, Inc.™ EP093

Animals of Europe

Hedgehog

Polecat

Alpine Ibex

Red Fox

Animals of North America

Bison

Canada Goose

Arctic Wolf

Polar Bear

Animals of South America

Llama

Two-Toed Sloth

Howler Monkey

Anteater